P9-ELO-246

Great Ideas of Science

PLATE TECTONICS

by Rebecca L. Johnson

Twenty-First Century Books
Minneapolis

For my brother Dan, the great scientist in my life

Text copyright © 2006 by Rebecca L. Johnson

All rights reserved. International copyright secured. No part of this book may be reproduced, stored in a retrieval system, or transmitted in any form or by any means—electronic, mechanical, photocopying, recording, or otherwise—without the prior written permission of Twenty-First Century Books, except for the inclusion of brief quotations in an acknowledged review.

Twenty-First Century Books
A division of Lerner Publishing Group
241 First Avenue North
Minneapolis, Minnesota 55401 U.S.A.

Website address: www.lernerbooks.com

Library of Congress Cataloging-in-Publication Data

Johnson, Rebecca L.
Plate tectonics / by Rebecca L. Johnson.
p. cm. — (Great ideas of science)
Includes bibliographical references and index.
ISBN-13: 978–0–8225–3056–5 (lib. bdg. : alk. paper)
ISBN-10: 0–8225–3056-2 (lib. bdg. : alk. paper)
1. Plate tectonics—Juvenile literature. I. Title. II. Series.
QE511.4.J64 2006
551.1'36—dc22 2004019732

Manufactured in the United States of America
1 2 3 4 5 6 – BP – 11 10 09 08 07 06

CONTENTS

INTRODUCTION

Frozen Fossils at the Bottom of the World

The frigid Antarctic wind howled as the scientists worked, bundled up in parkas, mittens, and heavy boots. They were perched on a cliff called Graphite Peak, located deep in the Transantarctic Mountains. Suddenly, one of the scientists shouted. He'd found something! The pale sun glinted off fossil bones nestled among the rocks.

As it turned out, the bones belonged to *Lystrosaurus,* an ancient reptile about the size of a big dog. It was 1969, and these were among the first fossils of a vertebrate animal (one with a backbone) ever found in Antarctica. What made this discovery even more remarkable was that *Lystrosaurus* fossils had previously been found in Africa, India, and China, where the reptile was known to have lived between 180 and 225 million years ago. This new find raised a puzzling question. How did *Lystrosaurus* get to Antarctica, an isolated continent at the bottom of the world?

The mystery went deeper. Fossils of ferns and trees had been found throughout the Transantarctic Mountains and in other parts of Antarctica. Delicate ferns don't grow in subzero temperatures. Obviously, Antarctica had been a much warmer place in the distant past. How had it transformed from a land of ferns and forests into a frozen wasteland?

The solution to these fossil mysteries is that Antarctica moved. Long, long ago it was joined to Africa, South America, and Australia. It lay closer to the equator then, where the climate is mild. And *Lystrosaurus*—together with other animals that later became extinct—roamed freely on this huge landmass.

Scientists found this *Lystrosaurus* fossil in South Africa.

Over time, the land broke apart and the pieces drifted away from each other. They spread all across the globe. The large piece that became Antarctica headed south. This drifting took millions of years. As Antarctica neared its current location, it slowly changed from a warm place with many plants and animals to a cold one with permanent snow and ice. The process that caused Antarctica—and all the other continents—to drift over time is known as plate tectonics.

Understanding plate tectonics starts with knowing Earth's basic structure. If you could slice through our planet, you'd find it is made up of three major layers. The innermost layer is a very hot core of iron and nickel. The inside of this core is solid, and the outside is liquid. The middle layer, the mantle, is composed of rock that flows very, very slowly, like toothpaste. The outermost layer is the crust. Oceanic crust forms the ocean floor, while continental crust forms the continents.

The crust and the upper part of the mantle, which is cooler and more rigid than the mantle's deeper parts, together make up what geologists call the lithosphere. The lithosphere is broken up into more than a dozen huge rocky slabs called tectonic plates. These plates are slowly moving. They ride along on top of the asthenosphere, a hot, semisolid part of the mantle that lies directly beneath the lithosphere.

The theory of plate tectonics is a relatively new scientific concept, developed and accepted by scientists in the 1960s. It has become a fundamental theory in geology, the science that deals with Earth's structure and processes.

Cross Section of the Earth

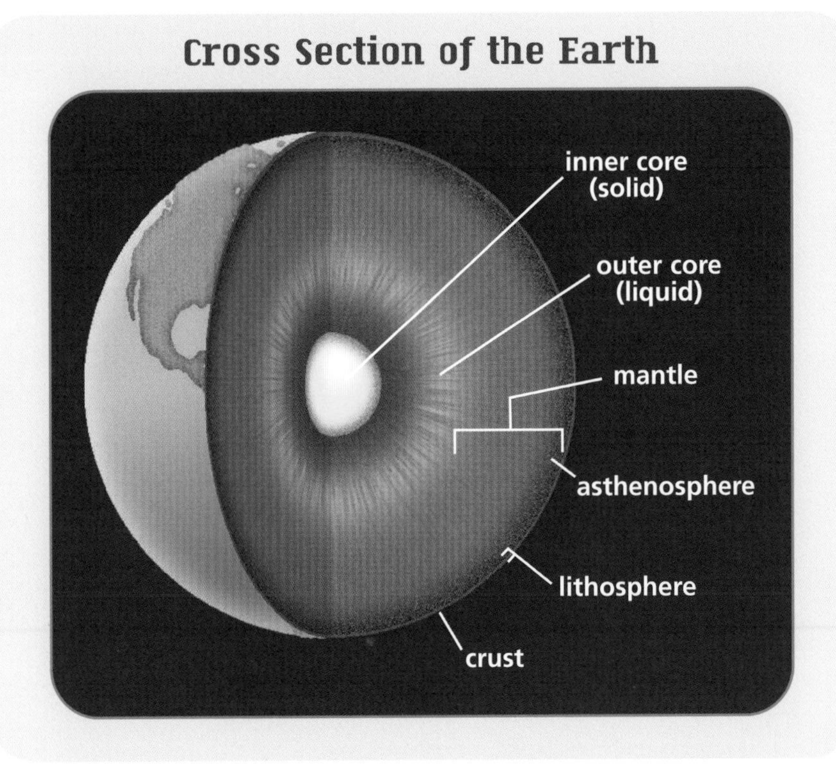

Plate tectonics not only explains the mystery of *Lystrosaurus* and drifting continents, it also describes how Earth's features have come to exist and how they continue to change.

How scientists arrived at the theory of plate tectonics is a story of daring ideas, heated controversy, and unexpected discoveries. It started with a seemingly outlandish idea that was eventually proven to be true, thanks to the creative detective work of many scientists. When the theory of plate tectonics emerged, it completely revolutionized our view of Earth and how it works.

CHAPTER 1

A PERFECTLY PUZZLING FIT

Take a look at South America and Africa on a map or globe. Do you notice anything interesting about the shapes of these two continents? If you could slide them together across the Atlantic Ocean, they would fit together almost perfectly, like two pieces of a jigsaw puzzle.

More than four hundred years ago, when the first maps of the entire world were drawn up, observant people noticed the same thing. The bulge on the eastern side of South America tucks neatly into the notch in Africa's western coast.

In 1596 the Dutch mapmaker Abraham Ortelius commented on this amazing alignment in a book he wrote on geography. To Ortelius, it looked as if South America and Africa had once been joined but were later torn apart. In 1620 English philosopher Francis Bacon wrote that the apparent fit of the two continents didn't look to be an accident. But Bacon didn't offer any explanation as to when or how South America and Africa might have been joined or separated.

Bacon may have had good reason for not offering an explanation. At the time, the European view of the world was dominated by religion and a very literal interpretation of the Bible. People who presented ideas that contradicted church doctrine could be fined, imprisoned, or even executed.

According to church officials, there was only one explanation for how the continents and oceans were arranged. They had been created that way. It was thought that Earth formed from a ball of intensely hot, molten material. As the young planet cooled, continents appeared on its surface—complete with mountains, valleys, and rivers—surrounded by oceans. Any changes to Earth's features that had occurred since creation were attributed to rare violent catastrophes, particularly the flood described in the Bible's book of Genesis. The German theologian Theodor Lilienthal proposed that the flood was so powerful and devastating that it had torn some of the continents apart. He used the nearly perfect fit of South America and Africa as proof.

Lilienthal and others who believed that Earth was changed only by rare—and rapid—catastrophes were called catastrophists. They were supported by the calculations of James Ussher, a seventeenth-century archbishop of the Church of England. Ussher maintained that Earth was only about six thousand years old. He reached this conclusion after developing a "biblical chronology" that put the date of creation at October 23, 4004 B.C. The catastrophists reasoned that since Earth was only six thousand years old, changes to its surface features had to have happened quite quickly.

Much More than a Guess **In science, the word *theory* has a very specific meaning. If you say you have a theory about how your dog got out of the yard, what you really mean is that you can make an educated guess about how he escaped. A better word to use might be hypothesis, which is essentially an educated guess, or a suggested solution to a problem based on some observation or information.**

When trying to solve problems in science, researchers often begin with a hypothesis. They may test whether or not that hypothesis is valid by conducting an experiment or making more observations.

A theory, on the other hand, is much more than an educated guess. It is a thorough and well-tested explanation for a fundamental principle or broad group of observations in science. Theories are typically supported by a great deal of evidence.

Ussher's dates were accepted for nearly two hundred years, despite much evidence to the contrary—including records of family histories from China that extended back before Ussher's date of creation! Belief in a six-thousand-year-old Earth, and the catastrophic view of how it had changed, endured until the early 1800s. By then evidence was building that Earth was much older than six thousand years and that its surface had changed slowly and steadily.

Learning from Rocks

Much of this evidence came from rocks. Scientists noticed how rain, wind, and waves gradually wear away, or erode, rocks. These forces of nature gradually grind rocks down, producing tiny particles of sand and silt. Wind and rain carry these eroded particles into rivers, lakes, and oceans where they settle on the bottom to form sediment. As time passes, new layers of sediment are added on top of older ones.

Geologists also noticed that some exposed cliffs and mountainsides are made up of different rock layers. These layers appear to have been laid down over very long periods of time. And like sediment at the bottom of bodies of water, younger layers are on top of older ones. In some places, people noticed that rock layers are bent or twisted. Other layers contain fossils of extinct plants and animals. Especially puzzling were the fossils of extinct sea creatures found in the middle of continents or at the tops of tall mountains—far from any existing ocean.

James Hutton

One of the first people to synthesize these observations into a hypothesis was James Hutton. He proposed a new view of Earth's age and changing features. Hutton had trained in Scotland as a medical doctor, but he was fascinated by geology. He read everything he could find about rocks, fossils, and Earth's history. Hutton traveled all over

Europe, studying soils, sediments, the effects of erosion, rock formations, and other geological features.

Everything Hutton saw on his travels suggested Earth was much more than six thousand years old. On one of his walks, for example, Hutton came upon an old stone wall that had been built in A.D. 122. Although the wall was more than 1,600 years old, the stones from which it was made were only slightly eroded. This impressed Hutton. Centuries of erosion had changed the stones in the wall only slightly. Surely, he reasoned, it must take far longer for jagged mountain peaks to become rounded hills and for rivers to cut deep valleys into the landscape.

Furthermore, Hutton discovered that many geological features are made of rocks formed from layers of sediment, like those deposited in bodies of water. He suggested that long ago, these sedimentary rocks began as tiny eroded particles that were washed off the land and deposited in the ocean. New layers were deposited on top of older ones. Gradually, over time, these layers of sediment were transformed by heat and pressure into hard sedimentary rock.

Hutton went on to suggest that in some parts of the world, ancient seabeds had somehow been raised up to become dry land. This idea explained why fossils of ancient ocean animals appeared on mountaintops. Hutton suggested that new land was also formed by volcanic eruptions and other processes that brought material from deep inside Earth to the surface.

Based on his extensive observations, Hutton concluded that changes in Earth's features could be explained entirely by the slow, steady work of natural forces. No sudden catastrophes or biblical explanations were necessary. Erosion wore rocks away. The fine particles produced by erosion were deposited to form sediment. The sediment was transformed into rock. Powerful forces raised sedimentary rock up out of the water or brought molten rock up from Earth's interior to create dry land. As soon as new land was formed, it started to erode, beginning the cycle all over again.

Hutton found evidence in his travels that Earth's surface had eroded and then been rebuilt by these same natural processes over and over again. His observations led him to conclude that Earth was many millions of years old and that the forces that had shaped Earth's surface in the past are still at work in the present.

Hutton made these daring ideas public when he presented several scientific papers to the Royal Society of Edinburgh in 1785. But he was not a good speaker, and the audience didn't understand his arguments. Ten years later, Hutton fleshed out his ideas and backed them up with extensive evidence in a book, *The Theory of the Earth.* Unfortunately, Hutton's writing style was just as difficult to understand as his speeches had been. When he died in 1797, the significance of his work had not yet been appreciated.

A few years later, John Playfair, one of Hutton's friends and a professor at the University of Edinburgh in Scotland, clearly summarized Hutton's ideas in the book *Illustrations of the Huttonian Theory of the Earth.* Suddenly, people understood Hutton's revolutionary proposition that Earth's surface changes uniformly, not catastrophically, over long periods of time.

Hutton's theory marked a turning point in the field of geology. The catastrophic view of a changing Earth began to fade. It was eventually rejected entirely by most geologists thanks to Charles Lyell, a British naturalist. In 1830 Lyell published *Principles of Geology* to support and expand on Hutton's theory. This new view of Earth came to be called uniformitarianism because it stressed uniform change over time. Geologists acknowledged that catastrophes, such as earthquakes and volcanic eruptions, still have a role in shaping our planet, but they attributed much less significance to this role. Uniformitarianism formed the foundation of modern geology and helped answer many geological questions.

But what about the continents and their near-perfect fit? Had they once been joined and somehow drifted apart? Uniformitarianism didn't offer a solution to that mystery. While erosion and other natural forces could alter some features, no one believed they could break up entire continents and shuffle them across the face of the planet.

Charles Darwin's Reading List

Lyell's writings shaped the thinking of many nineteenth-century naturalists, including Charles Darwin. When Darwin boarded HMS *Beagle* in 1831, he brought along a copy of Lyell's *Principles of Geology*. During the *Beagle*'s five-year voyage around the world, Darwin read and pondered Lyell's book. Darwin eventually applied the ideas of uniform change over time to living things. Darwin's theory, called natural selection, revolutionized the science of biology.

Chapter 2

From Uniformitarianism to Continental Drift

Uniformitarianism helped explain how Earth has changed over time. But since it failed to answer the question about the puzzling fit of the continents, scientists continued the search for clues that might solve this mystery. Some of the most interesting clues they uncovered came from rocks found in far-flung parts of the world.

About the time Hutton's ideas were becoming known, German explorer and naturalist Alexander von Humboldt was traveling in South America. On his voyage, von Humboldt came across rock formations along the Brazilian coast that were almost identical to rock formations known from the coast of western Africa. These matching rocks seemed to indicate that the continents had once been joined.

Von Humboldt speculated that South America and Africa had once been part of the same landmass. At some point in Earth's history, von Humboldt suggested, a huge mass of moving water had carved out a gigantic valley that became the Atlantic Ocean. It left eastern South America and western Africa with parallel coastlines that looked as if they had once fit together. Von Humboldt's suggestion was an interesting one, but there was no proof of such large-scale water action.

As time passed, other naturalists and explorers unearthed additional evidence that suggested the continents had been connected. They discovered, for instance, that some of the same kinds, or species, of turtles, snakes, and lizards living in South America are also found in Africa. North America and Europe have living things in common too, such as grizzly bears (known as brown bears in Europe), wolverines, and trees such as beeches and larches.

Other clues came from fossils. For example, camel fossils were discovered in North and South America. Yet living camels are found only in Africa, the Middle East, and parts of Asia. The preserved remains of other forms of ancient life, such as a fossil fern called *Glossopteris,* were remarkably similar in Africa, South America, and several other continents as well. Each new discovery seemed to indicate that the existing continents had once been joined and then become separated. But the burning question remained: how?

Far-Fetched Explanations

In 1858 a rather dramatic explanation came from Antonio Snider-Pellegrini, an American writer living in Paris. Like his contemporaries, Snider-Pellegrini believed Earth began as a hot ball of molten rock. He proposed that as Earth slowly cooled and a crust formed on its surface, most of the material that would become the continents

Fossil Finds in the Southern Hemisphere

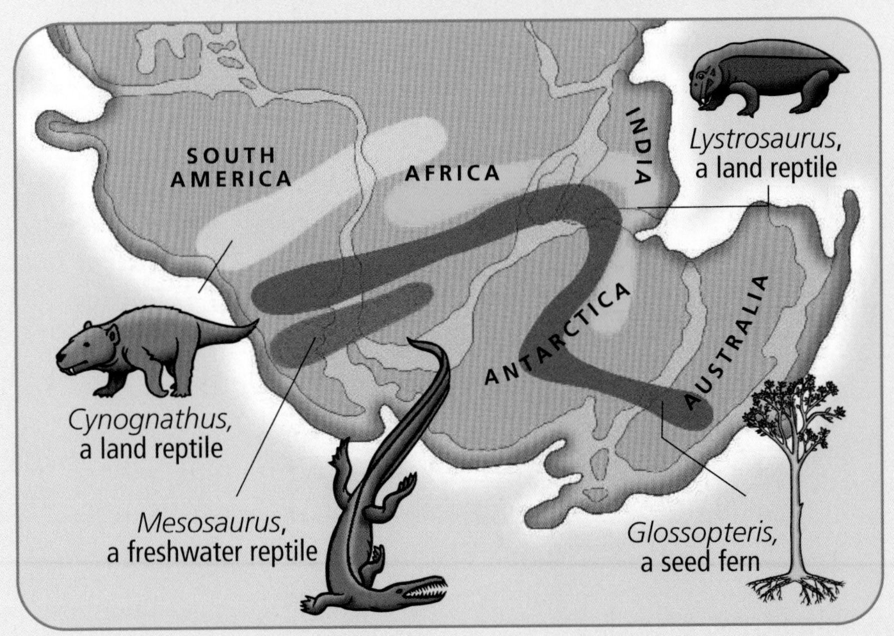

Locations of fossil remains of ancient plants and animals suggest the continents were once connected.

clumped together to form a single landmass on one side of the globe. This made the Earth unbalanced. Stressed by this lopsidedness, the landmass began to crack. Hot lava welled up through the cracks, and raging water began to push some of the broken pieces to the other side of the globe, forming North and South America and the Atlantic Ocean between them. Snider-Pellegrini cited the similarities of fossils found in coal beds and matching rock formations on opposing sides of the ocean as supporting evidence for his ideas. He even published a map that showed how the Americas, Europe, and Africa might have once fit together, like pieces of a jigsaw puzzle.

A few years later, George Darwin—the son of the famous naturalist Charles Darwin—suggested a different explanation. Darwin proposed that as the young Earth was cooling, it began rotating so fast that a huge chunk of the planet broke off and spun out into space. That chunk became the Moon. The great hole left behind began to fill with lava. The flow of material into this hole dragged what are now the Americas away from Europe and Africa. Eventually, the lava cooled and water filled the remaining depression, creating an ocean.

The ideas proposed by Snider-Pellegrini and Darwin about how the continents may have separated weren't widely accepted. They relied too much on global catastrophes for which there was no real evidence.

Snider-Pellegrini's maps from the mid-1800s show how he thought the continents had once been joined.

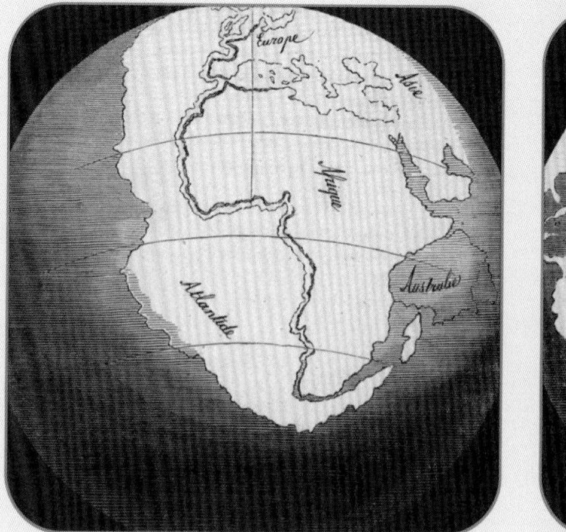

Ocean Investigations

While naturalists of the 1800s were probing remote corners of Earth's continents, other researchers were undertaking the first large-scale investigations into the nature of the ocean environment. In the 1850s, the U.S. Navy began taking depth measurements of the Atlantic Ocean. The measurements were needed to lay a telegraph cable between Europe and America. Up until this time, most people, including most earth scientists, assumed that the ocean floor was a relatively flat, featureless plain, shallow at the edges and deepest in the middle, like a soup bowl. They also expected it to be covered with a thick layer of sediment formed from the remains of countless sea creatures and dirt washed off the land over millions of years.

But the navy's survey of the Atlantic had unexpected results. The middle of the Atlantic Ocean was not its deepest part. Instead, the seafloor rose up in the middle, forming what seemed to be a high ridge roughly equidistant from North America and Europe. This high spot was first nicknamed Telegraph Plateau. Later, it came to be called the Mid-Atlantic Ridge.

In the 1870s, ocean exploration continued when HMS *Challenger* set out on a four-year mission to gather information about the ocean floor. Scientists on the ship collected data from more than 350 sites. They took nearly five hundred soundings (depth measurements) and dredged up material from the bottom in more than 130 different locations. They measured the distance to the bottom by throwing lead-weighted lines over the side of the ship and measuring out the line until the weight hit bottom. The researchers also measured currents and water temperature and made detailed charts showing the contours of the ocean floor. The scientists discovered that the ocean floor was a complex landscape, covered with hills and mountains and cut by great valleys, some of which were many miles deep.

As more people came to accept uniformitarianism, any explanation that involved catastrophic events was usually dismissed. However, the ideas of Snider-Pellegrini and George Darwin were important because they suggested the continents had once been joined as a single landmass that later broke apart.

Land Bridges

Several other scientists tried to explain the presence of similar rocks and fossils on widely separated continents in an entirely different way. They suggested that animals and plants had moved from continent to continent across bridges of land. They hypothesized that land bridges had once connected all the world's continents and then later had sunk below the surface of the sea.

By the late 1800s, naturalists had discovered enough plant and animal fossils around the world to lend strong support to the idea that land bridges had once linked South America, Africa, India, Australia, and Antarctica. One of the strongest advocates of the land bridge hypothesis was Austrian geologist Edward Suess. Like many scientists of his time, Suess believed Earth began as a molten ball that slowly cooled and formed a crust. Suess proposed that as Earth continued to cool, the crust contracted and shrank like a plump grape drying into a shriveled raisin. As a result of this shrinking process, some parts of the crust rose up to create mountain ranges while other parts sank downward, forming ocean basins.

In 1885 Suess published the first volume of *The Face of the Earth*. In the book, Suess proposed that up until about 225 million years ago, there was one enormous continent in the Northern Hemisphere and a similar supercontinent in the Southern Hemisphere. According to Suess, the northern landmass, Atlantis, consisted of the present-day continents of Europe, Asia, and North America. The southern supercontinent, Gondwanaland, included present-day South America, Africa, India, Australia, and Antarctica. A body of water he dubbed the Tethys Sea separated the two large landmasses.

A Tectonic Legacy **Edward Suess's *The Face of the Earth* was quite an undertaking. He wrote this four-volume explanation of the geology of the entire Earth from 1883 to 1909. Suess named his southern supercontinent Gondwanaland after the fossil-rich Gondwana region in India.**

Suess suggested that over time, parts of the supercontinents sank and were covered by water. For a while, the newly forming continents were still connected by a system of wide land bridges. These bridges made it possible for living things to move freely from one continent to another. Eventually, as Earth continued to cool and contract, the land bridges were completely covered by water.

The idea of land bridges seemed reasonable to many people. But try as they might, ocean explorers weren't able to find any evidence of sunken land bridges stretching between Africa and South America or Europe and North America. The problem became more complicated as new fossil and geologic finds were made. Not only were similar fossils found on widely separated continents but certain kinds of fossil remains revealed that the climate had changed in many parts of the world. For example, fossils of trees and ferns were found on lands near the North and South Poles, indicating that those regions had once enjoyed much warmer climates. Deserts and other hot places near the equator showed signs of having once been home to glaciers and animals that lived in cooler, more temperate climates.

As the evidence piled up, the confusion deepened. What could account for all these discoveries and observations? The explanations proposed during the 1800s all fell short in one way or another. Catastrophes were a return to the past. No evidence was found that sunken land bridges spanning major oceans ever existed. And no one seemed able to explain how the climate in some parts of the world could have changed so dramatically over time.

A Radical Idea

In 1908 two American geologists independently proposed that the continents *had actually moved* during past ages. Frank Taylor hypothesized that about 100 million years ago, two large continents existed, one at the North Pole and the other at the South Pole. He proposed that the Moon was much closer to Earth then—so close that it exerted a gravitational pull on the two continents, tugging them slowly toward the equator. As the landmasses moved, they broke into smaller pieces. As these pieces plowed across the ocean floor, the leading edges of some

The Alaska Connection No land bridges ever spanned the Atlantic or Pacific Oceans. But evidence suggests that a land bridge did connect northern Siberia to Alaska across the Bering Sea. Time and again, that bridge disappeared when sea levels rose and then reappeared when sea levels fell. Scientists think the Bering Sea land bridge last rose above the waterline about seventy thousand years ago. It connected Asia and North America until about eleven thousand years ago, when sea levels rose high enough to submerge it once again.

may have wrinkled up. Or perhaps they collided to create mountain ranges such as the Andes in South America and the Himalayas in Asia.

Howard Baker had a somewhat similar suggestion. But instead of the Moon, he thought Venus had come close to Earth and set the continents in motion. Taylor and Baker were among the first to propose that the continents had traveled across the face of the planet to their present positions.

Neither Baker nor Taylor could offer any concrete evidence to support his hypothesis, however. When Taylor presented his ideas in a paper to the Geological Society of America, they were largely dismissed as nonsense. How could continents made of solid rock move? Such an idea seemed outrageous to most scientists of the time.

CHAPTER 3

ALFRED WEGENER'S CONTROVERSIAL HYPOTHESIS

Alfred Wegener had a passion for many areas of science, especially meteorology, glaciology, and Earth science. He taught at the University of Marburg in Germany and had a gift for seeing tough problems—and their solutions—clearly, often in a flash of insight.

In 1911 Wegener was browsing in the university's library. By chance he picked up a scientific paper that proposed a land bridge had once existed between South America and Africa. The paper mentioned that identical fossils of plants and animals were found on both continents. Wegener was instantly intrigued.

Eager to learn more, Wegener tracked down other papers written about matching fossils and land bridges. The more he learned, the more he wanted to know. But the more Wegener considered the things he'd read, the more he became convinced land bridges could not explain matching fossils and rock formations on different continents. The problem with land bridges was that they depended on the widely accepted hypothesis that Earth had cooled and contracted, giving rise to mountains where the crust heaved up and ocean basins where it sank inward. According to this hypothesis, the land bridges that once connected the continents had been covered by water when the ocean basins filled.

Alfred Wegener *(left)* and a guide in Greenland in 1930

But Wegener reasoned that if the contraction hypothesis were correct, then mountains would be fairly evenly distributed over Earth's surface. They would also all be about the same age. But this wasn't the case. A glance at a world map shows that mountain ranges crop up rather haphazardly. The Andes Mountains run down the western edge of the South American continent, very close to the coast. The Himalayas lie deep in Asia's interior. In North America, the Appalachian Mountains are around 225 million years old, while the Rocky Mountains are only about 65 million years old.

Floating Continents

Assuming the contraction hypothesis was incorrect, Wegener puzzled over how else matching fossils and rock formations on widely separated continents could be explained. A clue came from Edward Suess. In addition to lending support to the idea of land bridges, Suess had written about discoveries that revealed the continents are made of relatively lightweight rocks, such as granite.

Much of the ocean floor, however, is composed of heavier, denser rock called basalt. According to the principle of buoyancy, lighter, less dense materials tend to rise above those that are heavier and denser. Ice, for example, is less dense than water. That's why ice cubes float in a glass of water. With the principle of buoyancy in mind, Suess proposed that the less dense continents "float" on the denser ocean rock beneath them.

Wegener applied the buoyancy principle to the hypothesis of land bridges. He believed if the land bridges had been forced down somehow, eventually they should rise again. The fact that they hadn't made him doubt that they had ever existed.

Furthermore, Wegener reasoned, if continents were able to move up and down, they should also be able to move sideways, drifting toward or away from each other. If this sort of "continental drift" was possible, it would solve the mystery of matching rock formations and identical fossils found in widely separated parts of the world.

Wegener realized that without proof, his hypothesis about drifting continents was no more convincing than land bridges or catastrophic floods that carved out ocean basins. So he set out to gather evidence to support his idea. He pored over maps, scientific papers, and reports—everything he could get his hands on.

In studying charts of world's mountain ranges, Wegener noticed that if he slid the continents together, mountain ranges on opposite continents matched up. They formed continuous ranges that spanned two or

REBOUNDING ROCKS **Edward Suess had found clear evidence that rock can "float" in northern Scandinavia. In places along the coast, people had noticed for hundreds of years that the land had been gradually rising. The reason? The entire region had once been covered by huge, heavy glaciers. When the glaciers melted about ten thousand years ago, the great weight left. Slowly but surely the land rose, or rebounded. Geologists call this process isostatic rebound. It is similar to what happens when a ship carrying a heavy cargo is unloaded—it floats higher in the water when the weight is removed.**

more landmasses. Wegener compared these matching mountain chains to torn bits of newspaper. He said, "It is just as if we were to refit the torn pieces of a newspaper by matching their edges and then check whether the lines of print ran smoothly across. If they do, there is nothing left but to conclude that the pieces were in fact joined in this way."

Wegener was convinced these matching mountain ranges showed that certain continents had once been joined. Matching rock formations and fossils also supported his case. Furthermore, his hypothesis explained why the climate had changed in some places. Deposits of coal in Antarctica and fossils trees on islands north of the

Arctic Circle indicated that those lands had once had a warmer climate. If those lands had once been near the equator and then moved slowly toward the poles, their climates would have changed over time too.

By the end of 1911, Wegener had gathered together an impressive amount of data in support of continental drift. Armed with this evidence, he spoke at two geological meetings in January 1912. In his speeches, which he titled "The Formation of the Major Features of the Earth's Crust," Wegener boldly suggested that all the landmasses on Earth had been joined together as a huge supercontinent. When the supercontinent broke apart, the pieces—the present-day continents—began drifting apart, slowly but steadily. Over time mountain chains were split, populations of plants and animals were divided, oceans formed as lands pulled apart, and climates changed as continents moved toward distant parts of the globe.

Bad Reactions

Wegener's extraordinary hypothesis was not well received by the scientists to whom he spoke. His ideas flew in the face of accepted geological ideas. And because Wegener had spent most of his career in the field of meteorology, not geology, he was regarded as an amateur, not an expert. Most of the scientists who heard Wegener's talks considered him an "outsider" who was attacking the foundations of geology.

Wegener wasn't discouraged by the negative reaction. He doubled his efforts to find more evidence in support of his ideas. In 1915 he presented his ideas and considerable evidence to support them in *The Origin of Continents and Oceans.* At first, Wegener's book didn't attract much attention outside Germany. World War I (1914–1918) was raging and the book was written in German. But in the years that followed, the war ended and Wegener revised *Continents and Oceans* several times. By 1924 it had been translated into English, French, Swedish, Spanish, and Russian. Suddenly, earth scientists all over the world were reading Wegener's theory of continental drift.

Wegener proposed that the huge supercontinent, which he now called Pangaea (all lands), had begun to break up about 200 million years earlier. North and South America drifted west, Asia went north, Australia east, and Antarctica south, while Europe and Africa remained relatively stationary. Wegener thought the continents moved through Earth's crust like icebergs plowing through the sheets of ice that cover the ocean's surface in polar regions.

As the continents drifted, plants and animals that had once shared the same landscape became separated. New oceans were formed. In some places, existing mountain ranges were pulled apart as continents separated. Where the crust crumpled and piled up along the leading edges of moving continents, new mountain ranges formed.

As the continents drifted, Wegener contended, some moved into different climatic zones. This explained why lands near the poles contain fossils of temperate and even tropical species. Wegener carefully plotted the worldwide distribution of many different fossils and rock types. His results indicated the former locations of tropical jungles, deserts, and ice caps. The history of climate change that emerged from this reconstruction made perfect sense if the continents had moved as Wegener proposed.

A Snail's Tale **While tracking down examples of similar or even identical animal and plant fossils on widely separated continents, Wegener learned about unlikely across-an-ocean relationships that existed between several living species. One of these was the common garden snail (*Helix pomatia*) that lives both in western Europe and eastern North America. Charles Darwin had suggested that perhaps some of the snails had been carried across the Atlantic Ocean on the feet of birds. But Wegener didn't think that was very likely. He believed a better explanation for the snail's distribution was continental drift.**

Wegener had a great deal of evidence to support his hypothesis. But explaining *how* continental drift occurred—the driving force behind it—was a bigger challenge. Here Wegener knew he was on shaky ground. He suggested two possibilities. The first had to do with centrifugal force—an outward pushing force created when objects are spinning. Imagine riding on a merry-go-round. As the merry-go-round spins, the force you feel pushing you outward is centrifugal force. Wegener hypothesized that perhaps the centrifugal force of Earth rotating on its axis had caused the continents to move. Alternatively, he proposed that the gravitational pull of the Sun and Moon might be enough to drag the continents across the face of the planet.

Continental Drift

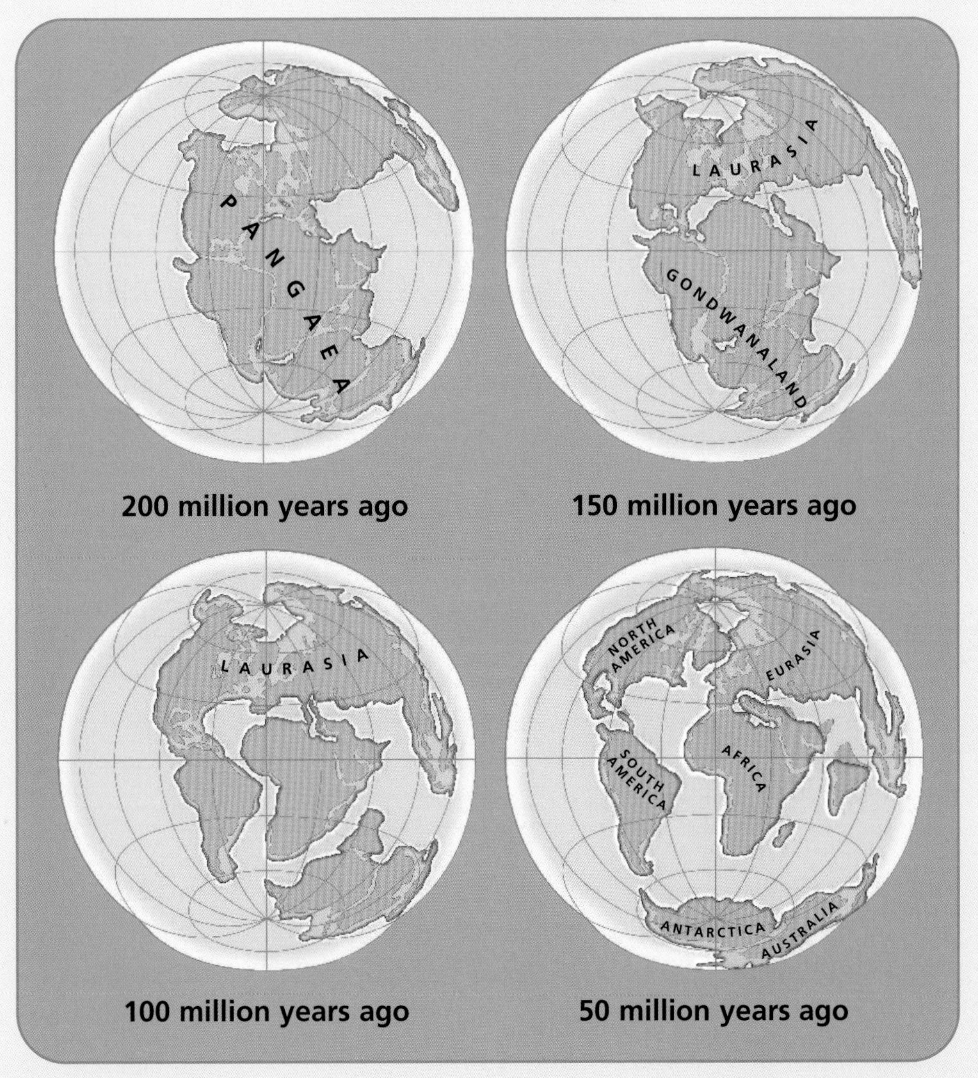

Alfred Wegener proposed that Earth's continents had once been joined as a single landmass, which he called Pangaea. This landmass gradually broke into pieces over many millions of years.

When scientists around the world read *Continents and Oceans,* most of them were outraged. Members of Britain's Royal Geographical Society completely rejected the hypothesis. The American Philosophical Society called it nonsense. It was dismissed as the stuff of fairy tales, or in slightly more scientific jargon, "geopoetry."

British astronomer and geophysicist Harold Jeffreys attacked the theory at its weakest point: Wegener's explanation for the driving force behind continental drift. With a few simple calculations, Jeffreys showed that a gravitational pull that was strong enough to shift continents would also stop Earth's rotation. According to Jeffreys, Wegener's hypothesis was impossible. No known force was powerful enough to move the lighter-weight continents through the heavier rock of the ocean floor.

Nevertheless, Wegener's theory of continental drift did find a few supporters. Wegener's strongest advocate was South African geologist Alexander du Toit. Du Toit had spent nearly twenty years investigating and mapping geologic features in South Africa. He had discovered dramatic similarities between plant and animal fossils he found there with fossils in South America. He felt that Wegener's hypothesis explained these similarities quite nicely. Du Toit even expanded on Wegener's ideas about Pangaea. He suggested there had been two supercontinents: Laurasia, formed by the present-day continents in the Northern Hemisphere, and Gondwanaland, made up of South America, Africa, Australia, India, and Antarctica.

Alfred Wegener's supporters were all but drowned out by his many critics. Yet throughout this storm of controversy, Wegener held steadfastly to his theory. As the debate raged, Wegener continued to gather more material in support of his controversial hypothesis. In the spring of 1930, he set out for Greenland to lead his third scientific expedition there. Bad weather plagued Wegner's team. After delivering supplies to a remote outpost, Wegener and a Greenlander guide began the return journey to Wegener's base camp on the coast. They never arrived. Months later, friends discovered Wegener's frozen body buried in the snow. They concluded Wegener had died of a heart attack and that his companion—who was never found—had buried him in the snow.

Wegener's death was reported in newspapers around the world. He was praised for his achievements as a meteorologist and polar explorer. However, his continental drift hypothesis seemed to die with him.

CHAPTER 4

UNLOCKING THE OCEAN'S SECRETS

Technology developed in the years leading up to the 1950s gave scientists new tools for exploring the ocean floor. Ocean scientists knew from depth surveys that the seafloor was a complex landscape, complete with mountains, valleys, and gorges. Two techniques—first echo sounding and, later, sonar—helped fill in more detail.

Echo sounders and sonar equipment use sound to measure the distance from the ocean's surface down to the ocean floor (or to an underwater object). A sound, usually described as a "ping," is broadcast down into the water from the ship. The time it takes for the sound to strike the ocean floor and bounce back to the ship—like an echo—is measured very precisely. Since the speed at which sound travels through water is known, the time it takes for the ping to travel down and back can be used to measure the distance to the ocean floor.

Echo sounders typically provide information only about water depth directly beneath a ship. Sonar is a little more complex. It provides information about distance and direction to objects, including the seafloor. In the 1950s, most ocean research ships were fitted with sonar. On the device's small screen, researchers could "see" the contours of the ocean floor over which the ship was passing. With the help of this remarkable tool, they began making very detailed maps of the ocean floor.

Sonar's Start **Sonar was originally designed to help ships spot icebergs in the water—before they ran into them. The U.S. military began using sonar to locate enemy submarines underwater during World War I (1914–1918) and World War II (1939–1945). After World War II, the U.S. government funded a huge effort to map the world's ocean floor. Soon hundreds of navy ships and oceanographic research ships were equipped with this new tool for ocean exploration.**

Sediment Secrets

Other scientists were investigating ocean floor sediments. The ocean floor was thought to be Earth's oldest feature. Most researchers expected the sediment covering it to be as much as 12 miles (19 kilometers) thick. They believed ocean floor sediment consisted of millions of years' worth of eroded soil and the remains of countless dead ocean organisms. In this sediment, scientists hoped to find fossil clues to Earth's history.

A technique called explosion seismology was developed to measure the thickness of ocean sediments. Explosion seismology involves lowering seismic recording devices down to the seafloor and then dropping explosives over the side of a ship. The explosives are detonated several hundred feet below the surface. When the shock waves from the explosion reach the ocean floor they are changed into seismic waves that traveled through ocean floor bedrock and the sediment overlying it. The seismic recording devices pick up and record these waves. The waves look slightly different depending on whether they travel through solid rock or much softer sediment. By carefully analyzing seismic wave records, researchers can tell how thick the sediment is in the part of the ocean they are studying.

Maurice Ewing, a scientist working with the Woods Hole Oceanographic Institution in Massachussetts, was an explosion seismology expert. In 1947 Ewing received a grant to investigate ocean-floor sediments using explosion seismology and by bringing up core samples—long, cylindrical "plugs" of sediment from the bottom of the ocean.

Maurice Ewing

Ewing and his team headed out into the Atlantic Ocean aboard the research ship *Atlantis.* What they found came as a complete surprise. The ocean sediment they measured was far thinner than expected. The deepest layer, located just beyond the shallow waters that surround the continents, was only a few thousand feet thick. Ewing calculated that such a layer would be, at most, about 200 million years old. But Earth was roughly 4.6 billion years old. Why was the sediment so thin and so young?

A Pretty Old Planet

Scientists believe our planet is approximately 4.54 billion years old. But they can't determine Earth's exact age because our planet's earliest rock has eroded or been destroyed by natural processes. If there are any "original" rocks on Earth from the planet's very beginning, they haven't been discovered yet.

Rocks older than 3.5 billion years are found on all of the world's continents. The oldest known Earth rocks, found near Canada's Great Slave Lake, are just over 4 billion years old. Scientists calculate the age of rocks by measuring the radioactivity of certain elements within them. These elements "decay" at a known rate. That is, they change from one form to another in ways that allow scientists to calculate when the rock was formed. This technique is called radiometric dating.

So how do scientists know Earth is just under 4.6 billion years old? They've assumed that our planet is the same age as the other planets in our solar system. Rocks from the Moon are 4.4 to 4.5 billion years old. The oldest meteorites—fragments of asteroids that formed when the solar system formed—are 4.53 to 4.58 billion years old. Taking into consideration several other factors, the calculations for Earth's age (and the solar system's) result in a date of 4.54 billion years, give or take a few million years.

Then came an even bigger surprise. Near the Mid-Atlantic Ridge, seismic tests showed almost no sediment at all. And, intriguingly, rocks dredged up from the bottom showed the signs of having been heated to high temperatures. Some rocks were actually rounded globules of pillow lava—volcanic rock that forms when lava flows into water and cools very suddenly. Ewing and his team were intrigued by these findings. It appeared that the ocean floor was much younger than anyone had imagined. And it had been formed by volcanic action.

Making a Map

Over the next few years, Ewing went to sea again and again, taking more measurements and conducting more seismic tests. In the early 1950s, Ewing decided to gather echo soundings of the ocean bottom that had been taken by ships sailing all over the Atlantic Ocean and combine them to make a detailed map that showed the contours of a large area of ocean floor. Bruce Heezen, one of Ewing's graduate students, and Marie Tharp, a cartographer working at the Lamont Geological Observatory in New York City, took charge of this huge task. Gradually, the shape of the underwater landscape in the North Atlantic came into focus on paper.

The Mid-Atlantic Ridge was one of the map's most prominent features. But rather than being an isolated high spot in the Atlantic Ocean, the Mid-Atlantic Ridge turned out to be a long, wide mountain range that ran right down the middle of the entire Atlantic Ocean.

Interestingly, it ran almost exactly parallel to the coastlines of the continents on either side.

As more echo soundings were plotted, Tharp noticed something else. A deep crack, or rift, ran right down the center of the Mid-Atlantic Ridge, all along its length. This rift was puzzling. What could cause such a rift to form in an undersea mountain range?

Then Heezen asked another scientist at Lamont to plot the location of all recent undersea earthquakes that had been recorded in the Atlantic on Tharp's map. Lo and behold, the dots marking the earthquakes lined up along the rift that ran through the middle of the Mid-Atlantic Ridge.

Intrigued, Heezen and Ewing gathered information about the locations of mid-ocean earthquakes from all over the world. When they plotted these data on a world map, they discovered that "lines" of earthquakes ran through the other major oceans of the world, not just the Atlantic. From this evidence, the scientists guessed that there were mid-ocean ridges, also split by deep rifts, in all the major oceans of the world. Other earthquakes were concentrated where the scientists knew there were very deep valleys, or trenches, in the ocean floor.

A final bit of information made what was happening underwater even clearer. A British geophysicist, Edward Bullard, had recently used a special probe equipped with thermometers to measure heat in the oceans. He discovered that the temperature along the center of the Mid-Atlantic Ridge, near the rift, was significantly hotter than anywhere else on the ocean floor.

The World's Longest Mountain Range

The mid-ocean ridge system is an enormous submarine mountain chain. It runs down the middle of the Atlantic Ocean. From there the ridge plunges south, stretching almost to Antarctica, before curving east and then north into the Indian Ocean. There it splits into two branches. One branch runs up toward India and then angles further east toward the Red Sea. The other branch heads south and then jags east between Australia and Antarctica. As it approaches the coast of South America, the mid-ocean ridge angles north, ending in Mexico's Gulf of California. On average, the mid-ocean ridge towers about 15,000 feet (4,500 meters) above the ocean floor. If all the water were drained from the oceans, the ridge would be the most prominent and impressive feature on the face of the earth.

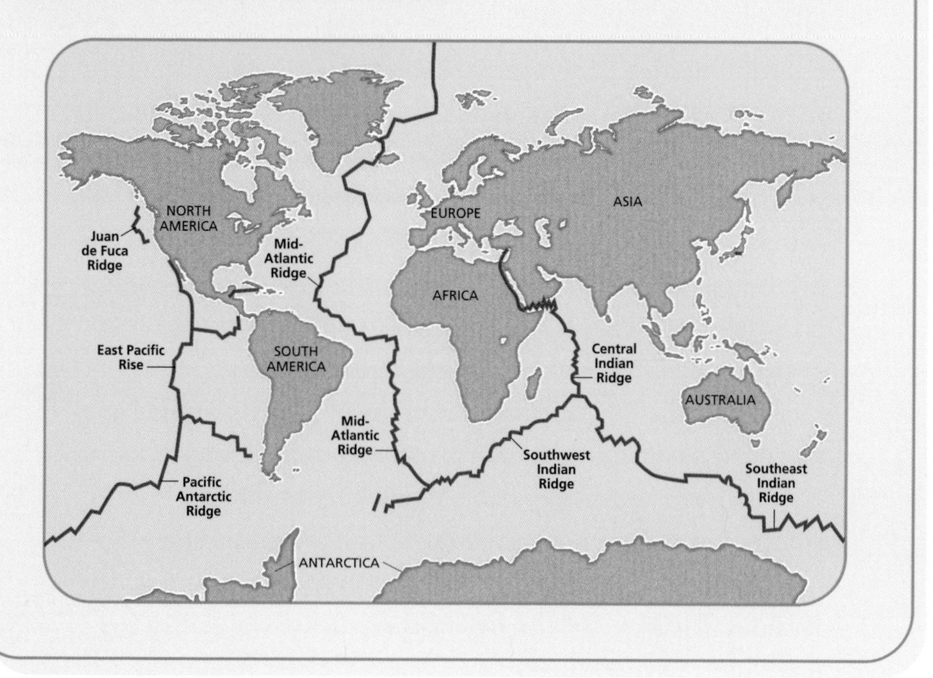

Adding this piece to the puzzle, Ewing, Heezen, and Tharp hypothesized that the Mid-Atlantic Ridge must lie over a great crack in Earth's crust. Molten rock from the mantle was coming up through this crack, causing earthquakes, forming the pillow lava that Ewing had dredged up, and raising the temperature of the water along the crack. The same was probably true for the other, yet-to-be-found mid-ocean ridges.

In 1959 the completed map of the North Atlantic seafloor showed an amazing mix of vast plains, cone-shaped volcanoes, and deep trenches, as well as the Mid-Atlantic Ridge. Over the next few years, this map of the ocean floor was expanded based on echo soundings, sonar readings, and other data gathered from all over the world. Heezen and Tharp had been correct in predicting the existence of other mid-ocean ridges. Ridges ran across parts of all the world's major oceans. And that's not all. The ridges were all connected. Together they formed a continuous undersea mountain range that zigzags among the continents and circles the globe like the stitching on a baseball. All along its length, the ridge is crossed at right angles by deep faults. To many geologists, these faults looked like the sort of fractures that form where great masses of rock are on the move.

Near many of the continents, deep trenches plunged into the ocean floor. Earthquakes were concentrated in these trenches, just as they were along the mid-ocean ridge. Something important was obviously happening along the ridges and at these incredibly deep spots on the ocean floor. More clues were needed to solve this submarine mystery.

Chapter 5

More Evidence

At about the same time that Maurice Ewing and his colleagues were making discoveries about seafloor sediments and the mid-ocean ridge, geologists on land were taking a new look at rocks. They discovered that some kinds of rocks had special magnetic properties.

If you've ever experimented with magnets, you know that a common bar magnet has two poles, a north end and a south end. If you take two magnets and push them together, their similar poles (a north and a north, or a south and a south) repel each other. But their opposite poles (a north and a south) attract each other. Furthermore, if you put a magnet under a thin piece of glass and then sprinkle iron filings on the glass, the filings line up along invisible lines of force that run between the two poles. These invisible lines of force make up the magnet's magnetic field.

Earth behaves as if it has a giant bar magnet inside it. It has both a north magnetic pole and a south magnetic pole. Earth's magnetic poles are near, but not at the geographic North and South Poles. The invisible lines of force that stretch between the north magnetic pole and the south magnetic pole form Earth's magnetic field. The poles exert a pull on the needle in a compass. The compass needle has two magnetic poles of equal strength at its ends: the north-seeking (N) pole and the south-seeking (S) pole, named for the directions they tend to point. When it's allowed to move freely, the north-seeking end of a compass needle points to the magnetic North Pole and the needle lines up along to the magnetic field lines that flow between the southern and northern magnetic poles.

Earth's Magnetic Mystery

Scientists don't know exactly what causes Earth's magnetic field. But the most widely accepted hypothesis has to do with currents within Earth's liquid core. This liquid core is made up mostly of iron and contains electrically charged particles. The movement of these charged particles inside Earth is thought to create the magnetic field.

Scientists had discovered that certain kinds of iron-rich rocks—especially volcanic basalt—have a faint magnetic field of their own. The scientists soon realized that the rocks' magnetic fields were acquired from Earth's magnetic field when the rocks formed. How did this happen? The iron-rich minerals in volcanic basalts act like tiny compass needles. When the rock is still liquid, these minerals line up with Earth's magnetic field, with their north-seeking ends pointed toward the magnetic North Pole and their south-seeking ends pointed toward the magnetic South Pole. As the rock cools and hardens, the orientation of these tiny mineral compasses is "locked in." They form a permanent record of the location of Earth's magnetic poles at the time when the rock formed.

Back in Alfred Wegener's day, scientists knew about this phenomenon, but it was almost impossible to measure. That changed in the 1950s when British physicist PMS Blackett invented the astatic magnetometer, which could detect extremely weak magnetic fields. The science of paleomagnetism (ancient magnetism) was born.

Blackett and other paleomagnetists used magnetometers to analyze rocks they collected from many sites across Europe and North America. The scientists were careful to gather rocks of different ages, from those that had been formed fairly recently in geologic time to much older rocks that had formed as much as 200 million years ago.

As the researchers analyzed their data, they discovered something quite odd. The magnetic orientation of the rocks they'd collected—the direction in which the fossil compasses pointed—differed noticeably from the current position of the magnetic poles. There were two possible explanations for this. Either the poles had moved or the landmasses had.

Some researchers pointed out that Earth's magnetic poles do indeed move. This strange phenomenon is called polar wandering. Earth's magnetic field is not static, like the field around a bar magnet. As a result, magnetic poles are constantly in motion. Since the magnetic North Pole was first located in 1831, it has wandered northwest by about 300 miles (500 km).

Paleomagnetists realized that if polar wandering alone was the cause of strange orientations of the fossil compasses in the rock samples, then rocks of the same age, no matter where they had formed, should point to the same place. But they didn't. When the scientists plotted out the fossil compass data on maps, a very interesting pattern emerged. For each landmass, the rock record indicated that the magnetic pole had changed position in a smooth curve, or pole path, over time. This didn't look like random polar wandering at all. But what did the pattern mean?

The answer came when the paleomagnetists experimented with changing the position of the continents. They discovered that the pole paths for Europe and North America could be made to match by bringing the continents together. The readings from the fossil compasses

WANDERING POLES **If Earth were exactly like a bar magnet, its magnetic poles would stay put. But because Earth's magnetic field is produced by the constantly moving liquid outer core, the magnetic poles roam in unpredictable ways. In 1831 explorer and naval officer James Clark Ross was the first to find the north magnetic pole. At that time, it was located along the edge of the Boothia Peninsula in the high Canadian arctic. Since then the pole has been moving slowly northwest. Its present location is northwest of Ellef Ringnes Island, about 800 miles (1,300 km) from the geographic North Pole.**

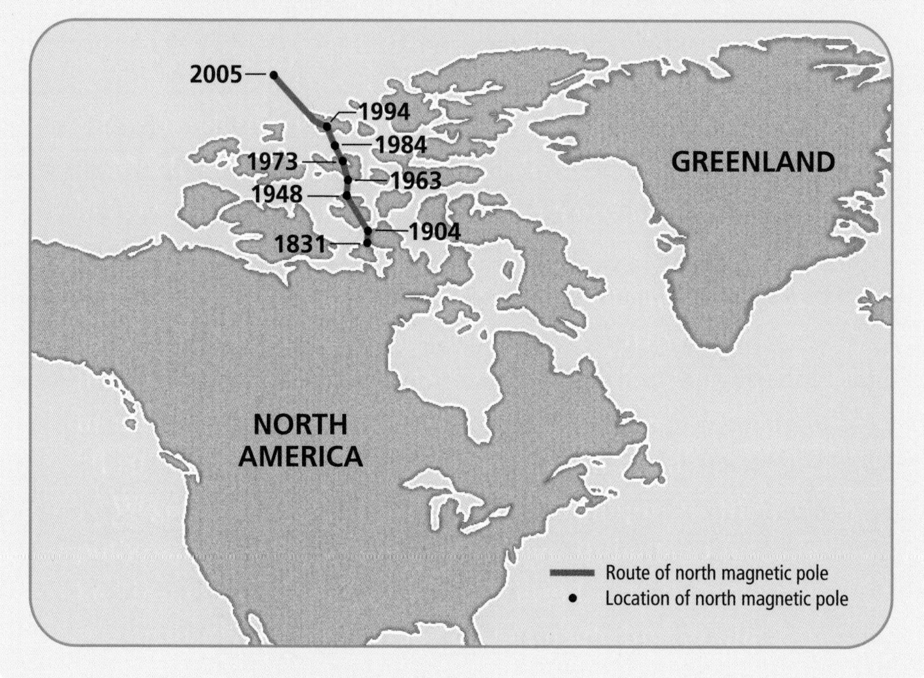

could be explained by assuming that the landmasses had once been joined together and then slowly drifted apart.

Scientists made more paleomagnetic comparisons of rocks in other parts of the world, such as Australia. The paleomagnetic evidence showed that continent had drifted and rotated its position over time. Paleomagnetic studies in India showed that it had once been located below the equator, in the Southern Hemisphere, and then slowly moved north to its current location.

The results of these paleomagnetic studies seemed to confirm what Alfred Wegener had suggested about thirty years earlier: that a supercontinent had split apart and the pieces—the modern continents—had slowly drifted to their current positions on the globe. A few scientists began to rethink the possibility of continental drift.

When North Is South

Earlier paleomagnetists had discovered something else about magnetism in rocks. Some rocks had what they called normal polarity. In other words, the north ends of the rock's many internal fossil compasses pointed toward Earth's magnetic North Pole. But other rocks had reversed polarity. In these rocks, the north ends of the fossil compasses pointed *south.* This second type of rock was evidence that Earth's magnetic field occasionally reverses polarity. The field flip-flops so that what was the north magnetic pole moves to the Southern Hemisphere and the south magnetic pole comes to lie in the Northern Hemisphere. Iron-rich rocks that contain fossil compasses

Arthur Raff and Ronald Mason charted the magnetism of the ocean floor near Washington state. The dark stripes on their map show areas of normal polarity on the floor of the Pacific Ocean. The light stripes indicate reversed polarity.

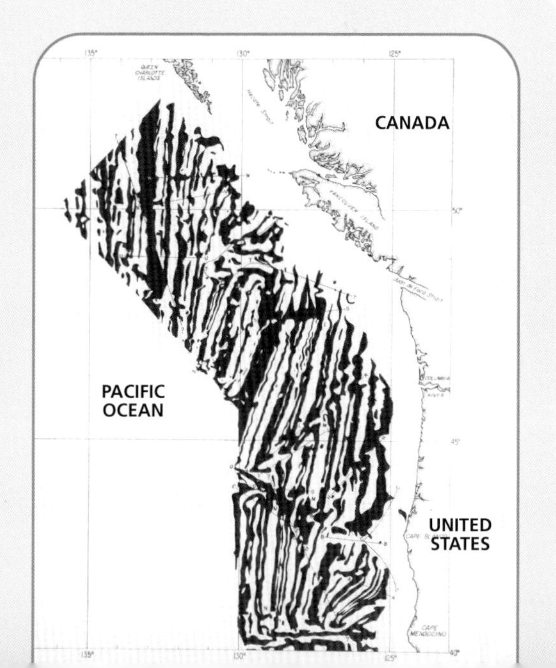

can show not only where the north and south magnetic poles were when the rocks were formed but also in which direction Earth's magnetic field was oriented.

In the early 1960s, scientists from the U.S. Geological Survey determined that Earth's magnetic field has reversed polarity 9 times during the past 3 million years. Other studies focused on older rocks and determined that 171 polarity reversals have taken place in the past 76 million years. No one knows exactly what causes these reversals. But when they occur, rocks forming at the same time record the event.

It was inevitable that interest in the magnetic fields of rocks would move from those on land to the rocks that make up the ocean floor. Using a torpedo-shaped magnetometer that could be towed by a ship, Arthur Raff and Ronald Mason set out to measure magnetic fields in the ocean floor off the coast of Oregon and Washington. When they charted their data onto a map, the results were very strange. Some parts of the ocean floor gave very strong magnetic readings. Others gave weak readings. What was stranger still was that these different areas alternated—strong, weak, strong, weak—across the ocean bottom. On a map, it looked like the ocean floor had stripes. The stripes ran almost parallel to the nearby coast. Raff and Mason had never seen anything like this magnetic striping before.

Hess's Hypothesis

On the other side of the North American continent, Princeton University geologist Harry Hess was quietly pondering many of the new findings. Hess was intrigued by what Ewing and others had discovered about the thinness of ocean floor sediments, especially that they were thinnest around the mid-ocean ridges. He mulled over the fact that a lot of volcanic activity occurred at the mid-ocean ridges. He also found it interesting that deep-sea trenches are the sites of much volcanic activity.

Harry Hess

Hess wove all these threads of evidence into a hypothesis about what was happening on the ocean floor. He proposed that the hot rock in Earth's mantle is slowly rising and melting beneath mid-ocean ridges, exerting pressure on the crust above it. The mid-ocean ridges are a crack in this crust. Along the crack, melted rock, or magma, continually oozes up from Earth's interior. As this hot lava hits cold seawater, it quickly cools to form new basaltic rock of the ocean floor. Inch by inch, this newly formed rock—new crust—spreads out on either side of the ridge. As more rock is added, it pushes older rock farther from the ridge.

This process takes place very slowly. But it means that the ocean floor acts like a conveyor belt, slowly but steadily moving away from the mid-ocean ridge. And riding on this rocky conveyor belt, Hess proposed, are the continents. As the ocean floor slowly spreads, the continents are carried along like groceries at the supermarket checkout.

Hess also reasoned that if new ocean crust is created along the mid-ocean ridges, then—since the planet is not getting bigger—old crust must be destroyed at the same rate. Hess hypothesized that old crust descends back down into the mantle in the deep-sea trenches. In short, the crust that makes up the ocean floor is constantly being recycled. New crust forms at the mid-ocean ridges. Over millions of years, it travels slowly away from the ridges until eventually it sinks back into the mantle at the deep-sea trenches.

Hess's hypothesis, which came to be known as seafloor spreading, made sense of many observations that had puzzled scientists for years. It explained the relatively thin sediment layer on the ocean floor. It also explained the relatively young rocks that composed it. None had been found that were more than 200 million years old. Most importantly, the hypothesis explained how continents could move. Hess concluded that Alfred Wegener had been right about continental drift. The continents had moved, and they were still moving. They didn't plow through the crust but were carried along on top of it.

But Harry Hess was keenly aware that using the phrase "continental drift" was likely to bring peals of laughter from most scientists around the world. Hess didn't present his ideas in a typical scientific paper. Instead, he described what he'd written using a term Wegener's critics had spouted years earlier: geopoetry. Hess's paper, titled "History of Ocean Basins," was widely read by scientists around the world. Most agreed that the idea of seafloor spreading was intriguing. But where was the proof? To propose an idea was one thing. To prove it was quite another.

Proving the Poetry

The proof was faster in coming than anyone could have guessed. In 1962, the same year that Harry Hess published his paper, British geologists Frederick Vine and Drummond Matthews were on a ship in the middle of the Indian Ocean. They were carrying out a magnetic survey of the ocean floor around the mid-ocean ridge.

When Vine and Matthews analyzed the results of their survey, they found that the area of ocean floor they were studying had the same striped pattern of magnetic readings that Mason and Raff had found in the eastern Pacific. As the two scientists pored over their data and ocean floor maps, they suddenly realized the strange stripes were a record of *reversals* in the polarity of Earth's magnetic field. Normal polarity, with the north magnetic pole in the Northern Hemisphere, produced a strong magnetic "stripe." Reversed polarity, with the north magnetic pole in the Southern Hemisphere, produced a weak magnetic "stripe."

This striped pattern of magnetic reversals on the ocean floor was strong evidence in support of seafloor spreading. As lava wells up from the mid-ocean ridges and becomes rocky crust, the newly formed rock records Earth's magnetic field at that moment in time. As more material is added, the new crust moves out on either side of the ridge. This forms a magnetic stripe that runs along each side of the ridge. When Earth's magnetic polarity reverses, new rock emerging at the ridge forms a different stripe (one with reversed polarity).

Over time, Vine and Matthews reasoned, periodic magnetic reversals should produce a series of alternating stripes that run parallel to the mid-ocean ridges. Furthermore, a magnetic stripe on one side of a mid-ocean ridge should have a mirror image on the other side. And this is exactly what they found. Their conclusion? The floor of the ocean does spread, just as Hess had hypothesized.

This discovery did more than support the seafloor spreading hypothesis. It also made it possible for scientists to measure how fast seafloor spreading was taking place. In studying rocks from many parts of the world, scientists had already calculated approximate dates for every polar reversal going back millions of years. That meant that the age of each stripe could be calculated.

Using this timetable of Earth's magnetic reversals, Vine and Canadian geophysicist J. Tuzo Wilson calculated the age of a number of individual magnetic stripes in certain parts of the ocean floor. By knowing the age of each stripe, they were able to calculate the rate at which new material was added to the ocean floor at the mid-ocean ridges. That, in turn, told them how fast the ocean floor was spreading—and how fast the overlying continents were being carried across the face of the globe.

Vine and Wilson calculated that the average rate of spreading was about 2 inches (5 centimeters) a year. It was more in some places, less in others. By using the average rate of spreading, the scientists determined that the Atlantic Ocean had opened up 150 million to 200 million years ago. Those dates corresponded perfectly with the ages of the oldest rocks dredged up from the bottom of the Atlantic. They were also remarkably close to the

Magnetic Polarity of the Ocean Floor

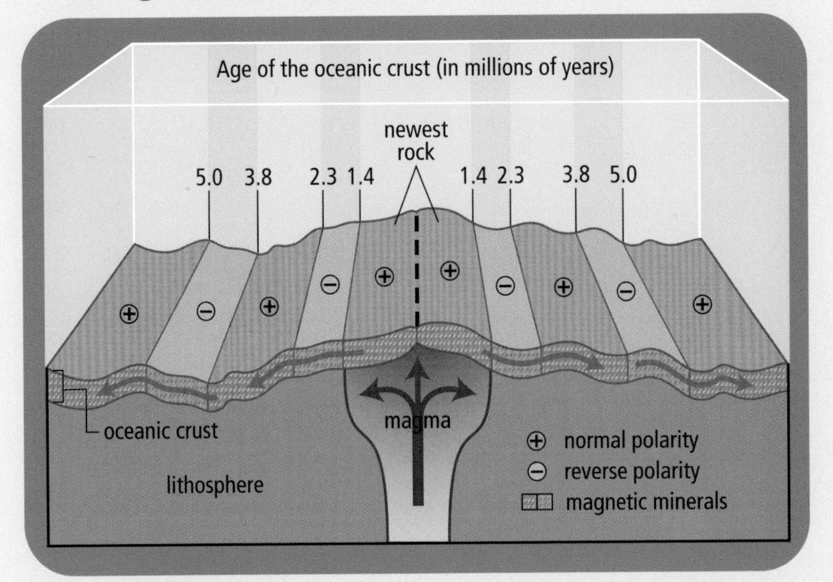

Magma rises from beneath the mid-ocean ridge *(center)* and emerges as lava from a rift at the center of that ridge *(dotted line).* The lava forces apart the oceanic crust on either side of the ridge, causing seafloor spreading. As the lava cools to form rock, its polarity is determined by the orientation of Earth's magnetic field. Changes in Earth's magnetic field result in "stripes" of normal polarity and reverse polarity along the ocean floor.

estimates Alfred Wegener had proposed a half century earlier for the formation of the Atlantic Ocean.

Paleomagnetic discoveries on land and on the ocean floor provided strong support for the seafloor-spreading hypothesis. Once-doubtful scientists began to examine the evidence more closely. Many were persuaded that

Hess and the others were onto something. More were convinced when additional evidence came from scientists who were studying earthquakes around the world.

In the early 1960s, a worldwide network of very sensitive seismographs—instruments that record movements in Earth's crust—was set up by the U.S. military. The network was used to monitor the world for explosions of nuclear weapons. Scientists who studied earthquakes, however, immediately realized that the seismograph network could also pinpoint the precise location of earthquakes occurring anywhere on the planet. As the data from the seismograph network poured in, a detailed map of earthquake sites emerged. As expected, most of the world's earthquakes were occurring along mid-ocean ridges and deep-sea trenches.

Seismologists used these results to show that the earthquakes occurring at mid-ocean ridges were relatively shallow. This finding tied in nicely with the assumption that Earth's crust was relatively thin at the ridge.

The earthquakes taking place in the deep-sea trenches, on the other hand, were located deeper in the Earth. This was strong evidence that ocean crust was descending into trenches and being recycled deep within Earth's mantle. As one part of the crust slid under another, the motion set off earthquakes. Taken together, paleomagnetic data from rocks and the earthquake readings from the worldwide network of seismographs persuaded many still-doubtful scientists that seafloor spreading does occur.

CHAPTER 6

A THEORY TAKES SHAPE

By the late 1960s, the ideas of seafloor spreading and continental drift were on their way to becoming widely accepted by many earth scientists. The stage was set for the formation of a theory to explain all the observations made on land and under the oceans.

In 1967 two young scientists working independently came up with the theory that is now known as plate tectonics. W. Jason Morgan of Princeton University in New Jersey and Dan McKenzie of Cambridge University in England proposed that Earth's outer 60 to 70 miles (97 to 113 km) was made up of a network of huge rock slabs that cover the entire surface of the planet like interlocking pieces of a jigsaw puzzle. The plates move very slowly in different directions and interact in different ways. Mid-ocean ridges, for example, are boundaries where two or more plates are moving away from each other. Ocean trenches are boundaries where one plate is sliding under another plate and heading down into Earth's interior.

Earth's Tectonic Plates

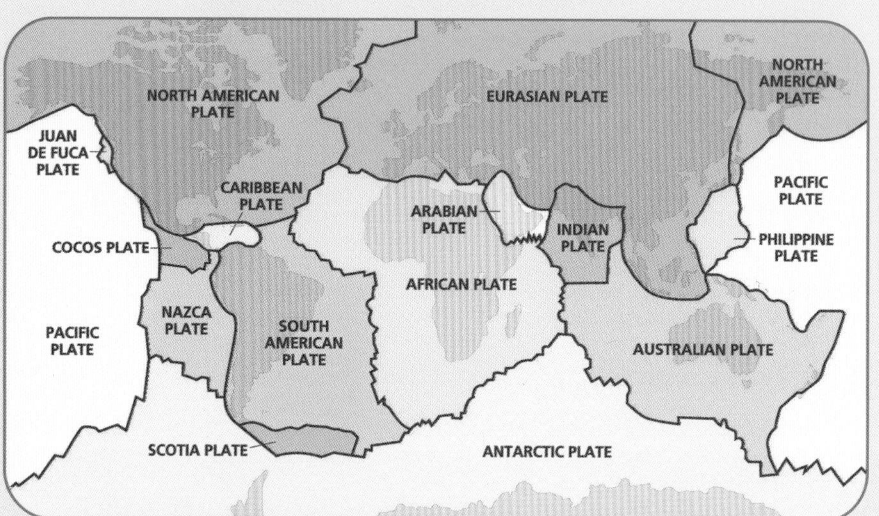

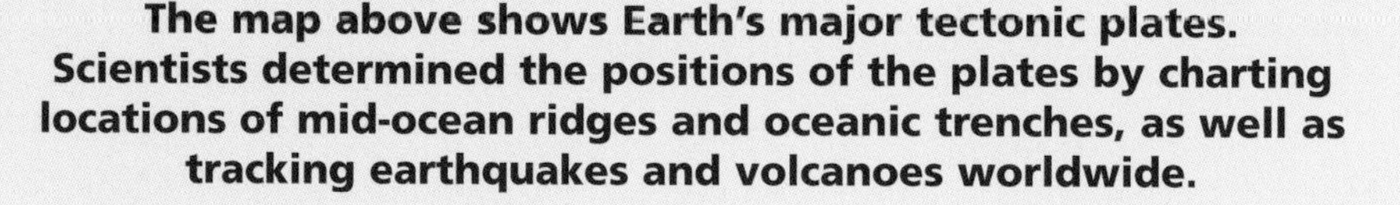

The map above shows Earth's major tectonic plates. Scientists determined the positions of the plates by charting locations of mid-ocean ridges and oceanic trenches, as well as tracking earthquakes and volcanoes worldwide.

The theory of plate tectonics quickly gained wide acceptance. Many skeptical researchers set out to try to disprove it. But in the end, their experiments provided even more evidence that the theory was valid.

Still, a few scientists charged that although the theory explained many observations, no concrete proof existed that the seafloor was moving away on either side of the mid-ocean ridges. That proof wasn't long in coming, however.

In 1968 the research ship *Glomar Challenger* set out on a yearlong expedition. It was equipped with a special drill, nearly 4 miles (6 km) long, that could take core samples from the ocean floor. Back in a laboratory, the cores were sliced down the middle lengthwise to reveal layers of sediment rich with the fossil skeletons of tiny ocean organisms. By dating the fossils and rock in the layers with radiometric-dating techniques, they were able to piece together a timeline of the site from which the core had been taken.

When researchers analyzed cores from either side of the mid-ocean ridge, they discovered that the seafloor becomes progressively older as the distance from the ridge increases. These findings provided conclusive proof that seafloor spreading does take place in the manner described by the theory of plate tectonics.

VOICES OF DISSENT

Not all scientists were fully supportive of seafloor spreading and the emerging theory of plate tectonics. One of the strongest opponents was Harold Jeffreys, who had found flaws in Alfred Wegener's continental drift hypothesis years earlier. Jeffreys didn't necessarily dispute the fact that plate tectonics was a real phenomenon. But he did criticize some of the explanations that were provided as to how it worked and what drove it.

UNDER THE SEA

By the early 1970s, scientists were finally able to witness firsthand what was going on at the mid-ocean ridges. In 1972 a French research ship towed a camera-carrying sled along the ridge near the Azores, a group of islands west of Portugal. Through the camera, scientists saw the same kind of rock—pillow lava—that Maurice Ewing had dredged up years earlier.

In 1977 scientists aboard the U.S. submersible *Alvin*—a small underwater vessel—made a remarkable discovery while exploring the rift in a mid-ocean ridge near the Galapagos Islands in the Pacific Ocean. At 8,500 feet (2,600 m) below the surface, the lights of the sub suddenly revealed something that looked like black smoke gushing

These tube worms, which can be 5 feet (1.5 m) long, live next to a hydrothermal vent in the Pacific Ocean.

from towers of volcanic rock. It wasn't smoke, but searing hot water black with volcanic chemicals. Even more amazing, the deep-sea hydrothermal (hot water) vent was packed with life. Giant tube worms swayed while ghostly white crabs picked their ways through fields of football-sized clams. This was scientists' first view of one of the strangest ecosystems on Earth, where life is fueled not by energy from sunlight, but by energy derived from the mix of chemicals surging out of the volcanic cracks. Billions of microbes in the water around the vents break down these chemicals and use the energy they contain to live and grow. The microbes, in turn, are food for many of the animals that live around the vent. In the years that followed, many other hydrothermal vent communities have been discovered along mid-ocean ridges all around the world.

How Plate Tectonics Works

After Morgan and McKenzie proposed the broad outline for the theory of plate tectonics, earth scientists around the world set to work investigating and testing its important details. They are still at it, fitting piece after piece into this exciting scientific puzzle. While many unanswered questions remain, the theory of plate tectonics has become a cornerstone of modern geology. It is the best explanation we have for how Earth has changed over time.

During Earth's long history, it has undergone many transformations. Much of this change is the result of moving tectonic plates. Tectonic plates existed early in our planet's history. They have been very slowly drifting ever since, repeatedly clustering together and then separating, changing the face and features of Earth.

The most dramatic evidence of change can be found at plate boundaries, places where two or more plates meet. Scientists give the plate boundaries different names based on the behavior of the plates.

In eastern Africa, three plates are moving away from each other, creating a divergent boundary *(above)* in the Great Rift Valley.

Divergent boundaries occur where plates are moving apart and new crust is being formed. The Mid-Atlantic Ridge is a good example of a divergent boundary. Far out in the middle of the Atlantic, beneath thousands of feet of seawater, several massive plates are slowly moving apart. At this seam, where the plates lie alongside one another, magma surges up from Earth's interior and cools to form new crust. Bit by bit, at the average rate of about 1 inch (2.5 centimeters) per year, the plates are spreading out, carrying the Americas slowly away from Africa and Europe and gradually widening the Atlantic Ocean.

Convergent boundaries occur where two or more plates are coming together or colliding. Where plates converge, one of several things may happen. If an oceanic plate meets a continental plate, the oceanic plate will slide under the continental plate. Oceanic plates always sink beneath continental plates because they are denser (and therefore heavier) than the continental plates. This process, called subduction, causes oceanic crust to descend into Earth's interior to become part of the mantle.

KILLER WAVES **Violent earthquakes that occur along subduction zones can trigger enormous ocean waves called tsunamis. These waves can travel across great distances at high speeds. When a tsunami reaches a coastline, it rises up to form towering waves that slam onto the shore, often causing terrible destruction. In December 2004, for example, a massive earthquake occurred near the Indonesian island of Sumatra. The quake generated a powerful tsunami in the Indian Ocean. It struck the coasts of Sumatra, Thailand, India, Sri Lanka, and several other countries. More than 150,000 people died and millions were left homeless, making it one of the worst natural disasters in recorded history.**

Off the western coast of South America, for example, the Nazca Plate is converging with and sliding below the South American Plate, which carries the continent of South America. The descending edge of the Nazca Plate drops down into a deep-sea trench that runs along the coastline of South America. At the same time, the western

Actions of Earth's Tectonic Plates

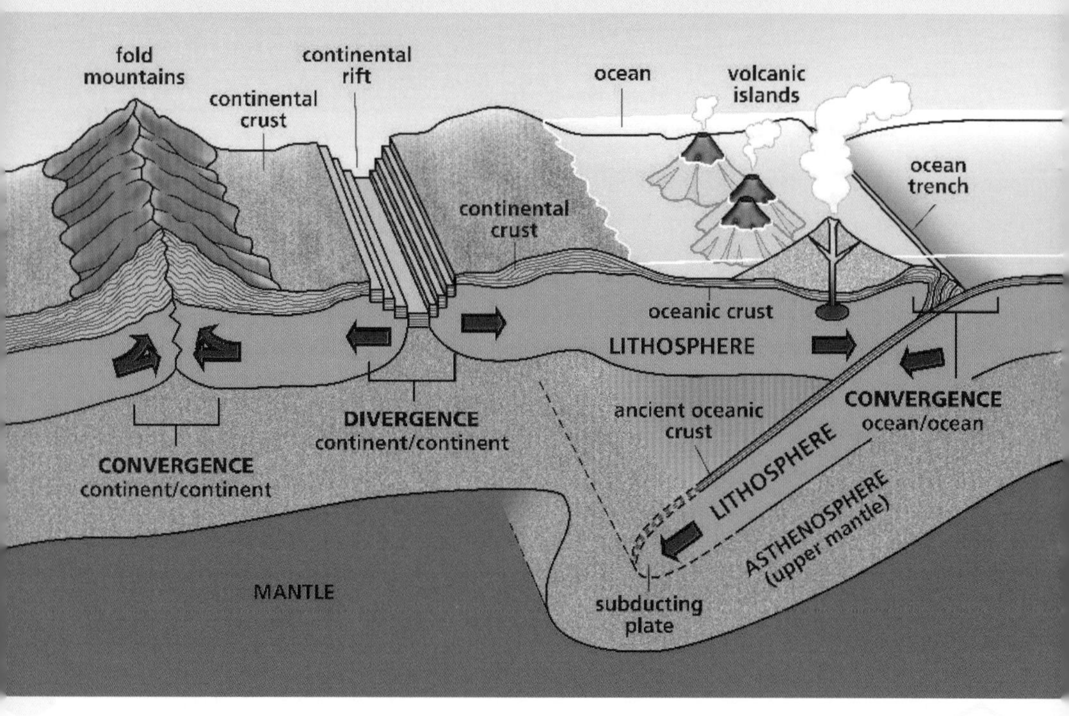

edge of the South American Plate is being lifted and crumpled up, a process that has helped create the towering Andes Mountains. Volcanoes and earthquakes are common all along this convergent boundary.

When two oceanic plates converge, the denser plate is subducted under the other. This forms a deep-sea trench. The deepest such trench in the world is the Marianas Trench in the western Pacific Ocean, where the Pacific Plate and the Philippine Plate converge. It is 36,000 feet (11,000 m) below the ocean's surface.

Subduction zones are the site of many active volcanoes. In some places, these volcanoes rise above sea level

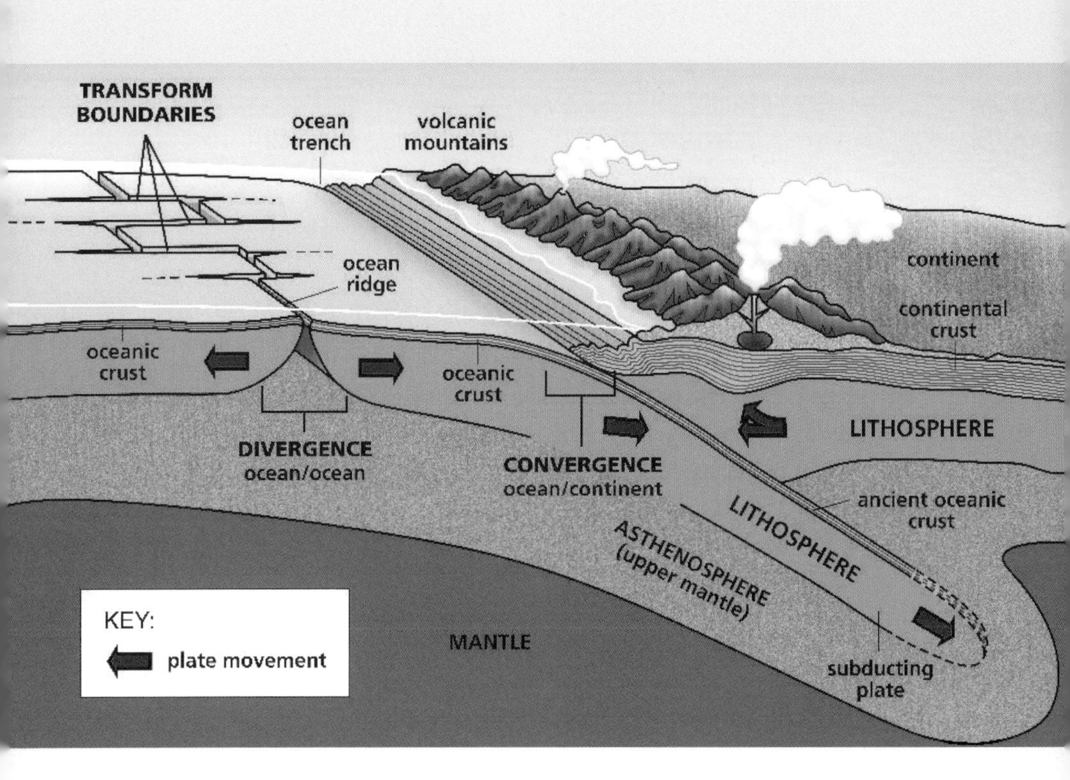

and form arc-shaped chains of volcanic islands. The Aleutian Islands off the coast of Alaska formed this way. Earthquakes occur regularly around these islands.

A third type of convergence produces some of the most spectacular and visible outcomes of plate tectonics. When two continental plates meet head on, neither one slides under the other. Instead they collide, causing the crust to break, buckle, and pile up into huge new mountain ranges, known as fold mountains.

About 225 million years ago, India was located south of the equator, near the continent of Australia. Slowly it began to move north, riding on the Indian Plate.

Eventually it crossed the equator and about 50 million years ago, it crashed into Asia, which rides atop the Eurasian Plate. As the two continents collided, their edges were crumpled and pushed up into what eventually became the Himalayas, the highest mountain range on Earth.

A third type of boundary between plates is a transform boundary. Here, plates slide past each other horizontally, forming large transform faults or fracture zones. Most of the world's transform faults are found on the ocean floor along the edges of the mid-ocean ridges. However, a few occur on land. One of the most famous is the San Andreas Fault zone in Southern California, where the Pacific Plate and the North American Plate slowly grind past each other. Many earthquakes occur along this fault.

The Great San Francisco Earthquake of 1906

Shortly before dawn on April 18, 1906, citizens of San Francisco, California, were jolted awake by a terrible earthquake. Violent shaking continued for nearly 45 seconds. All over the city, chimneys toppled and buildings collapsed. Streets cracked and buckled. On the Richter scale, used to measure the severity of earthquakes, the San Francisco quake was estimated to have been about 8.3, making it one of the most powerful earthquakes ever to hit North America. During the earthquake, the land on the east side of the San Andreas Fault moved southward—as much as 21 feet (6.3 meters)—relative to the land on the fault's west side.

A Hot Spot

Most volcanic eruptions take place near plate boundaries. But a few volcanoes form inside tectonic plates. The Hawaiian Islands are a good example. They are far from a plate boundary, yet they are volcanic. How do earth scientists explain this?

In 1963 J. Tuzo Wilson reasoned that perhaps these odd volcanic spots were the result of hot spots, or mantle plumes, in the mantle below the plates. Unlike the plates, these hotspots don't move. They remain in the same place while the plates moved along overhead. At a hot spot, Wilson suggested, a column, or plume, of hot rock rises from deep inside Earth. The plume partially melts right below the plate, causing undersea volcanoes—and ultimately volcanic islands—as the piled-up lava from countless eruptions eventually rises above sea level.

The hot spot theory caught on quickly. For years it went largely unchallenged. But in the late 1990s, a number of scientists actually went looking for mantle plumes, using high-tech seismology, in areas where they had been predicted to exist. They found some evidence of plumes under Iceland and Hawaii, but not at Yellowstone National Park in Wyoming. The details on mantle plumes remain unresolved. Expect the debate to continue until conclusive evidence is found to explain their origin.

The nature of hot spots is not the only unsolved question surrounding plate tectonics. The theory is, like most theories in science, a work in progress. As new information is discovered or new observations made, the theory is refined accordingly.

Convection and Gravity

One of the most fundamental unanswered questions about plate tectonics is proving a challenging one to answer. What force or forces provide the energy needed to move huge slabs of Earth's crust over the face of the globe?

Until the mid-1990s, most scientists thought that heat-driven motions, called convection currents, in the mantle were the most likely driving force behind plate tectonics. It was thought that the hot, fluidlike rock in the asthenosphere behaved like a big pot of thick soup being heated on a stove. Soup that is heated from below rises. When it reaches the top of the pot it moves outward, cools slightly, and then sinks downward along the sides of the pot. When it reaches the bottom, it's heated up again and the process repeats.

According to the convection hypothesis, hot mantle material supposedly moved upward toward the lithosphere, spread outward just underneath the tectonic plates, and then descended back into the mantle as it cooled slightly. The spreading of the mantle material was thought to pull and push the tectonic plates around on Earth's surface.

However, new technology has allowed scientists to learn more about Earth's interior. It turns out that the mantle doesn't really behave like a pot of soup. Its movements are far more complicated and its effect on tectonic plates may not be that significant.

So what's the answer? Most scientists now favor subduction and gravity as the primary driving forces behind plate tectonics. Remember that subduction affects only oceanic parts of the lithosphere. Denser or heavier oceanic plates sink beneath lighter oceanic or continental plates. Out in the ocean, where two oceanic plates meet, the older plate sinks. Why? Because as oceanic crust moves away from the mid-ocean ridges where it formed, it cools and grows thicker and denser. In time, the cold oceanic lithosphere becomes denser than the hot asthenosphere underneath. When that happens, there is a tendency for the edge of the colder, denser plate to begin to sink or subduct. It's a bit like a sheet of paper floating on water. Once one edge gets wet, it has a tendency to slide beneath the water's surface, pulling the whole sheet down with it.

Once a plate begins to subduct, gravity takes over. The edge of the descending plate continues to move downward, pulling the rest of the plate with it. Tectonic plates are often called "slabs," and the forces that cause them to move down into the mantle are known as "pull." Subduction, in the form of this gravitational slab pull, is now considered to be the largest driving force that moves massive tectonic plates across Earth's surface.

Scientists may debate the details of the process. But the theory of plate tectonics is almost universally accepted as the best explanation of the geological processes that affect Earth. The theory accounts for many observations that scientists have made about our planet over hundreds of years. Not the least of these is the "remarkable fit" of the continents, which sent the first geologists looking for answers.

Glossary

asthenosphere: a hot, semisolid part of the mantle that makes up the lower portion of Earth's upper mantle

buoyancy: the tendency of an object to float in liquid

catastrophic view: the belief that Earth is changed only by rare and rapid catastrophes

centrifugal force: an outward pushing force that acts on an object moving in a circular path

convergent boundary: the boundary between tectonic plates that are moving toward each other

core: the innermost layer of Earth, made of iron and nickel

crust: the outermost layer of Earth

divergent boundary: the boundary between tectonic plates that are moving apart from each other, or rifting

erosion: the gradual wearing away of land by ice, water, or wind

fold mountain: formed by the folding of Earth's crust along a convergent plate boundary

geology: the science that deals with Earth, its origins and evolution, its structure, and the processes that act on it

Gondwanaland: an early landmass that broke up to form India, Australia, Antarctica, Africa, and South America

island arc: an arc-shaped chain of volcanic islands that forms near an ocean trench

land bridge: narrow strips of land that connect two or more continents

Laurasia: an early landmass that broke up to form Asia, Europe, Greenland, and North America

lava: magma that has reached Earth's surface

lithosphere: a layer of solid rock that makes up the outer 62 miles (100 km) of Earth, including both the crust and the outermost portion of the mantle

magma: molten, or melted, rock beneath Earth's crust

mantle: the middle layer of the Earth—lying between the crust and the core

paleomagnetism: the study of natural magnetic traces within rocks that record the direction of Earth's magnetic field at the time that rock formed

Pangaea: a supercontinent that broke apart 200 million years ago to form Laurasia and Gondwanaland

polar wandering: the change in position of Earth's magnetic poles over time

reverse polarity: when Earth's north magnetic pole is located near the geographic South Pole and the south magnetic pole is located near the geographic North Pole

seafloor spreading: the idea that the ocean floor moves away from mid-ocean ridges over time

sediment: eroded particles of rock, in the form of sand and silt, that gather together at the bottoms of oceans, lakes, and other bodies of water

subduction: the sinking of an oceanic plate as the result of convergence with a less dense oceanic or continental plate

tectonic plates: large, rigid plates, or slabs of rock, that make up Earth's lithosphere

transform boundary: adjacent tectonic plates that are moving in opposite directions along a parallel line at their common edge

tsunamis: large ocean waves, often caused by underwater earthquakes or volcanic eruptions

uniformitarianism: the belief that geologic processes, such as erosion, that currently cause change in Earth's structure are the same processes that brought about change throughout Earth's history

volcanic mountains: peaks formed where magma erupts and piles up on Earth's surface, typically near subduction zones

Timeline

A.D. 79 Mount Vesuvius erupts, burying the Italian cities of Pompeii and Herculaneum and killing 20,000 people.

1492 Christopher Columbus makes his first voyage of discovery to the Americas.

1556 An earthquake near Shaanxi Province in China kills 830,000 people.

1596 Abraham Ortelius publishes *Thesaurus geographicus,* suggesting that South America and Africa had once been joined.

1658 Archbishop James Ussher concludes that Earth was created on October 23, 4004, B.C.

1789 George Washington becomes the first president of the United States.

1795 James Hutton publishes *The Theory of the Earth.*

1802 John Playfair publishes *Illustrations of the Huttonian Theory of the Earth.*

1811 Earthquakes in the midwestern United States temporarily change the course of the Mississippi River.

1815 Mount Tambora in Indonesia erupts, killing 92,000 people.

1830 Charles Lyell publishes *Principles of Geology.*

1831–1836 Charles Darwin travels around the world aboard HMS *Beagle.*

1850s The U.S. Navy begins taking depth measurements of the Atlantic Ocean.

1858 The first successful transatlantic telegraph cable is established.

1861–1865 The American Civil War is fought.

1885 Edward Suess publishes *The Face of the Earth.*

1905 Albert Einstein publishes his special theory of relativity, including $E=mc^2$.

1906 A major earthquake strikes San Francisco, California.

1908 Frank Taylor and Howard Baker independently propose that the continents move.

1914 World War I begins.

1915 Alfred Wegener publishes *The Origin of Continents and Oceans,* suggesting continental drift.

1918 World War I ends.

1939–1945 World War II is fought.

1959 Ewing, Heezen, and Tharp complete their map of the North Atlantic seafloor.

1960 The strongest earthquake ever recorded—9.5 on the Richter scale—hits the South American country of Chile.

1962 Harry Hess publishes his "geopoetry" essay.

1967 W. Jason Morgan and Dan McKenzie independently propose the theory of plate tectonics.

1969 *Lystrosaurus* fossils are discovered in Antarctica.

1980 Mount Saint Helens erupts in Washington state, killing 57 people.

2004 An earthquake off the coast of Sumatra generates a devastating tsunami in the Indian Ocean that kills more than 150,000 people.

Biographies

W. Maurice Ewing (1906–1974) Maurice "Doc" Ewing grew up on a farm in northern Texas. He became a physics professor, and the U.S. Coast Geodetic Survey enlisted him to study the ocean floor using seismic exploration methods. He made major discoveries about the continental shelf off the eastern coast of the United States. He went on to explore large parts of the ocean floor all around the world. With a team of researchers, he discovered the existence of the mid-ocean ridges and the seismic activity that takes place along them. Ewing's discoveries played a key role in proving the hypothesis of seafloor spreading and supporting the development of the theory of plate tectonics.

Harry Hammond Hess (1906–1969) Harry Hess was born in New York City. When he started at Yale University, he planned to become an electrical engineer. But after two years of college, he changed his major to geology. He became a professor of geology at Princeton in 1932. During World War II, Hess served in the navy, commanding an attack transport ship. His ship was equipped with echo sounding equipment—helpful both for spotting enemy submarines and for studying the ocean floor. In 1960 Hess developed what came to be called the seafloor-spreading hypothesis, which explained the relative young age of the ocean floor, the presence of island arcs, the deep-sea trenches, and the origin of the mid-ocean ridges. From 1962 until his death, Hess was chairman of NASA's Space Science Board. He helped plan the first landing of humans on the Moon. He died in August 1969, one month after astronauts aboard Apollo 11 walked on the Moon.

James Hutton (1726–1797) Born in Scotland, Hutton studied law and medicine before devoting himself to the natural sciences. He was a chemist, geologist, and naturalist. He is most famous for originating the idea of uniformitarianism, a fundamental principle of geology that explains the formation of Earth's features by steady, slow-acting natural processes over long periods of time.

Charles Lyell (1797–1875) Lyell was a British geologist, born in Kinnordy, Scotland, who spent much of his life traveling and popularizing scientific ideas. During his travels across Europe, Lyell discovered that different layers, or strata, in sedimentary rock found in widely separated areas could be identified by the particular marine fossils that they contained. The discovery of these "reference fossils" was important in helping to determine the geologic age of certain rock layers. However, Lyell is probably best known for his strong support of uniformitarianism.

Abraham Ortelius (1527–1598) Born in Antwerp, Belgium, Abraham Ortelius (also known as Abraham Ortel) is considered one of the most important and influential cartographers (mapmakers) of the sixteenth century. His maps were hand-colored and lavishly adorned with ships, mermaids, monsters, and other embellishments. In 1570 Ortelius compiled seventy of his maps, each engraved in a uniform size and supplemented with descriptive text, into a collection called *Theatrum orbis terrarium (Theater of the World). Theatrum* was the first comprehensive world atlas.

Edward Suess (1831–1914) An Austrian geologist, born in London, England, Edward Suess helped lay the foundation for plate tectonics. As a young man, Suess worked in the Natural History Museum in Vienna. He published several papers on marine fossils. Suess was intrigued by the fact that identical types of plant fossil deposits were found on widely separated continents. To explain this discovery, Suess proposed that all Earth's continents had once been linked into a single, immense continent that he called Gondwanaland. Suess wrote about his ideas in a four-volume book called *The Face of the Earth,* published between 1883 and 1909.

Marie Tharp (b. 1920) One of the few women involved in the development of plate tectonics, Marie Tharp drew the maps of the ocean floor that helped prove seafloor spreading and continental drift and led to the development of the theory of plate tectonics. A native of Ypsilanti, Michigan, Tharp learned about mapping from her father, a surveyor who made soil classification maps for the U.S. Department of Agriculture. She studied liberal arts in college. But in 1943, as many college-age men in the United States went off to war, several universities opened their doors to female students. Tharp earned an advanced degree in geology from the University of Michigan. She was hired as an assistant to Bruce Heezen at Columbia University in 1948. Using her mapmaking skills, Tharp plotted sonar data to create hand-drawn maps of the ocean floor. The maps revealed the mid-ocean ridges in startling detail.

Alfred Lothar Wegener (1880–1930) Born in Berlin, Germany, Alfred Wegener was the son of a minister who ran an orphanage. As a boy, he developed a passion for Greenland, a frozen landmass that would come to play a large role in his life. In 1905 Wegener received his doctorate degree in astronomy from the University of Berlin. But he turned his attention to meteorology and used kites and balloons to study the upper atmosphere. In 1906 Wegener and his brother flew in a hot air balloon for more than fifty-two hours, collecting atmospheric data. Their flight set a world record. Wegener is best known for developing the controversial hypothesis of continental drift and gathering an enormous amount of data to support it. But he also made three expeditions to Greenland, in 1906, 1912, and 1930. On those expeditions, Wegener had many perilous adventures and set several polar exploration records, including being one of the first people to spend the winter in Greenland. Wegener died during the 1930 expedition. His companions on the trip last saw him alive on the day of his fiftieth birthday.

JOHN TUZO WILSON (1908–1993) Canadian geologist and geophysicist J. Tuzo Wilson was born in Ottawa, Ontario, to adventurous parents who both loved mountaineering. Wilson inherited his parents' passion for exploration and the outdoors. At age seventeen, Wilson became a field assistant to mountaineer Noel Odell and in working for him, discovered the world of geology. Wilson went on to earn degrees in geology and physics from the University of Toronto, Cambridge University in England, and Princeton. At Princeton Wilson met Harry Hess, Maurice Ewing, and other scientists who would also play a major role in the development of the theory of plate tectonics. But up until the late 1950s, Wilson strongly supported the idea of a contracting Earth as an explanation for the formation of Earth's surface features. He opposed the idea of continental drift. At that point, evidence from the ocean floor and elsewhere convinced Wilson that he was wrong. He went on to become the world's leading spokesman in favor of continental drift and, ultimately, the theory of plate tectonics. It was Wilson who proposed that Earth is divided into several large, rigid plates that are moving apart as seafloors spread.

Source Notes

27 "On the Shoulders of Giants," EO Library: Alfred Wegener, n.d., http://earthobservatory.nasa.gov/Library/Giants/Wegener/wegener_4.html (October 15, 2004).

Select Bibliography

Golden, Frederic. *The Moving Continents.* New York: Charles Scribner's Sons, 1972.

Marvin, Ursula B. *Continental Drift: The Evolution of a Concept.* Washington D.C.: Smithsonian Institution Press, 1973.

Miller, Russell. *Continents in Collision.* Alexandria, VA: Time-Life Books, 1983.

Our Violent Earth. Washington D.C.: National Geographic Society, 1982.

Seyfert, Carl K., and Leslie A. Sirkin. *Earth History and Plate Tectonics: An Introduction to Historical Geology.* 2nd ed. New York: Harper and Row, 1979.

Sullivan, Walter. *Continents in Motion.* New York: McGraw-Hill, 1991.

Young, Patrick. *Drifting Continents, Shifting Seas.* New York: Watts, 1976.

Further Reading

Ballard, Robert. *Exploring Our Living Planet.* Washington D.C.: National Geographic Society, 1983.

Clark, John. *Earthquakes to Volcanoes.* New York: Gloucester Press, 1992.

Erickson, Jon. *Plate Tectonics: Unraveling the Mysteries of the Earth.* New York: Facts on File, 1992.

Kious, W. Jacquelyne, and Robert I. Tilling. *This Dynamic Earth: The Story of Plate Tectonics.* Washington, D.C.: U.S. Geological Survey, 1996.

Sattler, Helen Roney. *Our Patchwork Planet.* New York: Lothrop, Lee & Shepard Books, 1995.

Silverstein, Alvin, Virginia Silverstein, and Laura Silverstein Nunn. *Plate Tectonics.* Brookfield, CT: Twenty-First Century Books, 1998.

Websites

Continental Drift
http://www.oceansonline.com/continen.htm
This Web page discusses the history of continental drift and the development of the theory of plate tectonics.

Geologic Time
http://pubs.usgs.gov/gip/geotime/contents.html
This section of the U.S. Geological Survey website features articles that explain the concept of geologic time and how scientists have determined Earth's age.

Plate Tectonics Topic
http://scign.jpl.nasa.gov/learn/plate.htm
This NASA website includes an interactive diagram that describes Earth's structure.

This Dynamic Earth
http://pubs.usgs.gov/publications/text/dynamic.html
An Internet-based edition of the book *This Dynamic Earth,* this site includes many photographs and diagrams.

Index

Photo Acknowledgments

The images in this book are used with permission of:
© Jonathan Blair/CORBIS, p. 5; © Bettmann/CORBIS, p. 11; Courtesy University of California Library, Berkeley, California, p. 18 (both); © Alfred Wegener Institute, p. 25; © Woods Hole Oceanographic Institution, pp. 36, 58; © Todd Strand/ Independent Picture Service, Courtesy Geological Society of America/Arthur D. Raff/Ronald G. Mason, p. 47; Archives, Department of Geosciences, Princeton University, p. 49; © Altitude/Peter Arnold, Inc., 60.

Cover illustration by Tim Parlin/Independent Picture Service.